CW00500916

THE ILLUSTRATED MOTORCAR LEGENDS

ROLLS-ROYCE

ROY BACON

SUNBURST BOOKS

Acknowledgements

The author and publishers wish to acknowledge their debt to all who loaned material and photographs for this book. The bulk of the pictures came from the extensive archives of the National Motor Museum at Beaulieu, and we had kind assistance from Rolls-Royce itself while a picture or two came from the author's camera. Thanks to all who helped.

This edition published 1996 by Sunburst Books, Kiln House, 210 New Kings Road, London SW6 4NZ.

ISBN 1 85778 226 7

Printed and bound in China

CONTENTS

START OF THE LEGEND

1903-1906

ROLLS-ROYCE - 'The Best Car in the World' said the discreet and fashionable advertisements, a legend among cars from its earliest days, and the product of skills devoted to producing the very best, often regardless of cost or time. At 60mph the loudest noise in a Rolls-Royce came from the electric clock reported one road test; a Rolls-Royce engineer claimed it was no magic, simply patient attention to detail.

The Hon Charles Rolls was driving before the end of the 19th century and racing early in the 20th. To finance his passion he set up a business in London together with Claude Johnson to sell and service cars, most of which were imported, although by 1903 he was seeking a British replacement.

The start of the legend, a 1904 Royce with Charles Rolls at the wheel, a military passenger and servants for both.

One of the first fruits of the union of Rolls and Royce, the 10hp, two-cylinder car shown at the 1904 Paris Saloon.

Henry Royce was to fulfil Rolls's need for he was by then working seriously on the design of a car. An engineer, he had set up his company in Manchester to design and make electrical equipment which soon had the reputation for quality and reliability that was to characterise all his work. In 1902 he bought a French Decauville and soon established its weak points as things to avoid and its good ones as things to use, modify and improve on. He would continue this - as a good engineer does - using ideas from any source and adapting them to reach his high personal standards.

In 1904 Royce built three cars powered by a twin-cylinder, water-cooled engine. A mutual friend alerted Rolls to this and the quality of the cars Royce

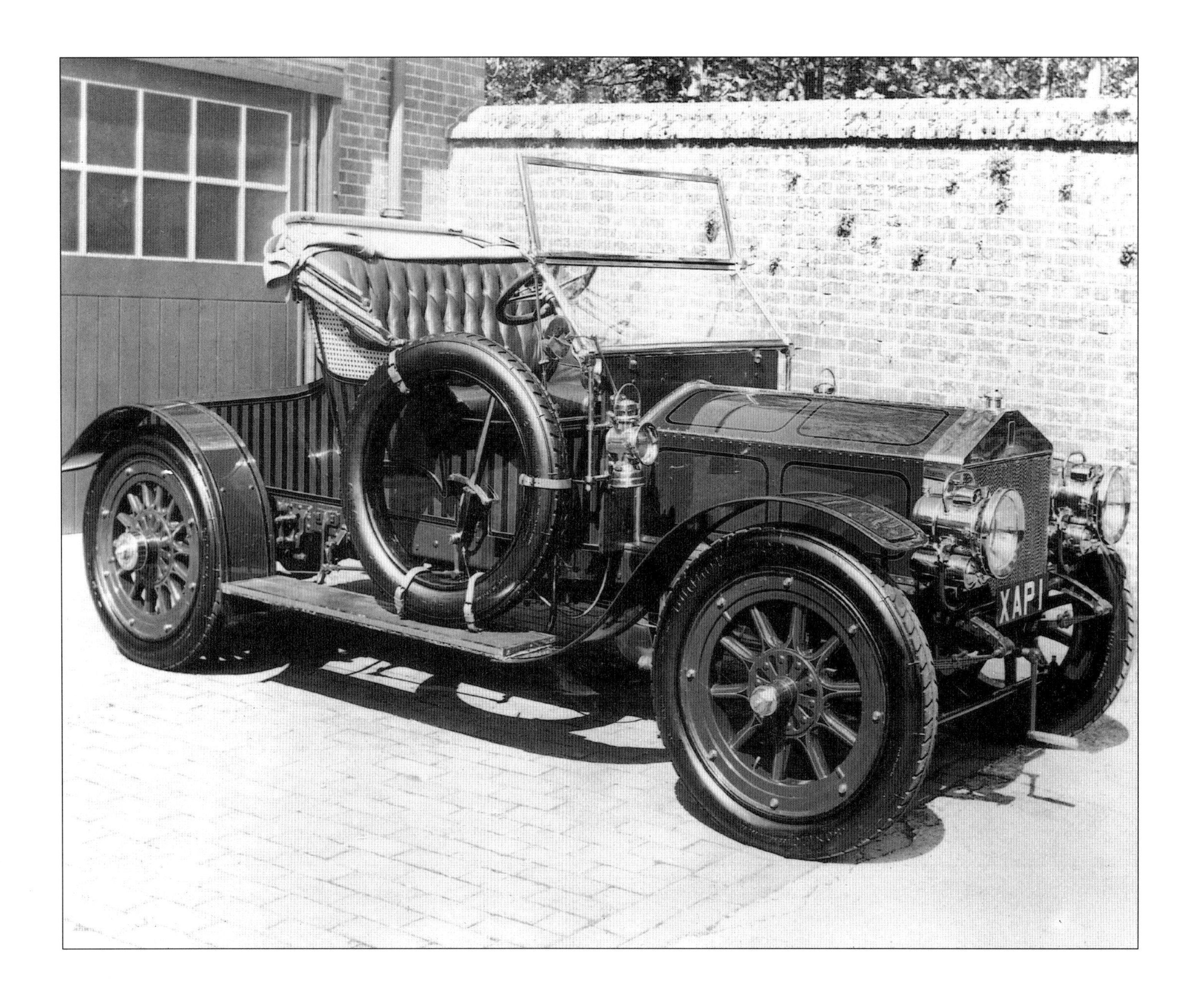

This is a 1906 30hp car whose chassis led to the Silver Ghost with a new design of six-cylinder engine.

was producing; Rolls went to Manchester and the outcome was an agreement for Royce exclusively to supply Rolls with a range of cars.

The range comprised the 10hp twin, a 15hp triple, 20hp four and 30hp six, for all of which Royce planned to have as many common parts as possible because the bore and stroke (100x127mm) was the same for all. The three had separate cylinders but the others had them cast as pairs and thus Royce showed his grasp of combining the needs of production with a reduction of the number of spare parts carried in the stores.

From the start the Rolls-Royce cars were built to high standards, with the emphasis being on a smooth and quiet ride, for Royce realised that noise would create antagonism against the car in those early days. The motors were conventional in layout, the engine water-cooled at the front driving back to the gearbox and thence by shaft to the rear axle. Braking was by a band on a drum behind the gearbox and expanding shoes for the rear drums, the wheels the wooden artillery type.

The chassis was built using channel steel and suspension was by semi-elliptic springs, those at the rear connecting to a transverse spring. Various bodies in the styles of the era were used. A feature from the start was cruise control by means of a governor that set the speed the car would maintain up or down hills, within the limit of the engine power.

These early cars were built in small numbers, some 17 twins in 1904-05 and six triples in 1905. Then the fours and sixes took over, around 40 of each being produced up to 1907 along with a trio of V-8s. The cars were all good, although not quite so much better than their contemporaries as Royce might have suggested. In addition, the length of the six-cylinder engines of that time, dictated by the early techniques used in building them, led to torsional crankshaft vibration and breakage. Thus, there was much argument as to the varying merits of the four- and six-cylinder engines.

While this went on the Hon Charles Rolls continued his racing, taking a car to the Isle of Man in 1905 for the Tourist Trophy race. It was the first race in which a Rolls-Royce started but it went out with gearbox trouble in the first two miles. However, Percy Northey was second in a 20hp model and Rolls returned in 1906 to win easily.

Meanwhile, Henry Royce was planning a new engine, refined and silent, to power a single new model that would replace all of the existing range.

In 1906 Charles Rolls won the Tourist Trophy race in the Isle of Man in this Rolls-Royce.

SILVER GHOST

1907-1925

THE car that became known world-wide as the Silver Ghost was listed as the 40/50 and its chassis was much as that of the Thirty that preceded it. It was the engine that set it apart from its past and the other great cars of the Edwardian times, and continued to do so well into the 1920s.

The new engine remained an in-line six but with two blocks of three cylinders mounted on an aluminium crankcase. This supported a substantial crankshaft in seven main bearings in its upper half in modern fashion, a considerable improvement on the older design. The bearing dimensions were such that a pressure lubrication system was essential but Royce knew about these, having used one in the V-8 engine.

The bore and stroke of the engine were equal and given as 4.5in, the capacity as 7,036cc and the compression ratio as 3.2:1. Power output was 48bhp at 1,250rpm, which was average for the time. Unlike the early engines that had overhead inlet valves, the 40/50 had side valves that allowed the cylinder heads to be cast in one with the block, to avoid leaking joints - then a common problem. The

Chassis number thirteen, the famous, fabulous Silver Ghost that gave the 40/50 its name and created the legend.

LEFT: One side of the six-cylinder Silver Ghost engine that set new standards for many a year.

BELOW: This early Silver Ghost was attending a Rolls-Royce rally at Beaulieu in 1960.

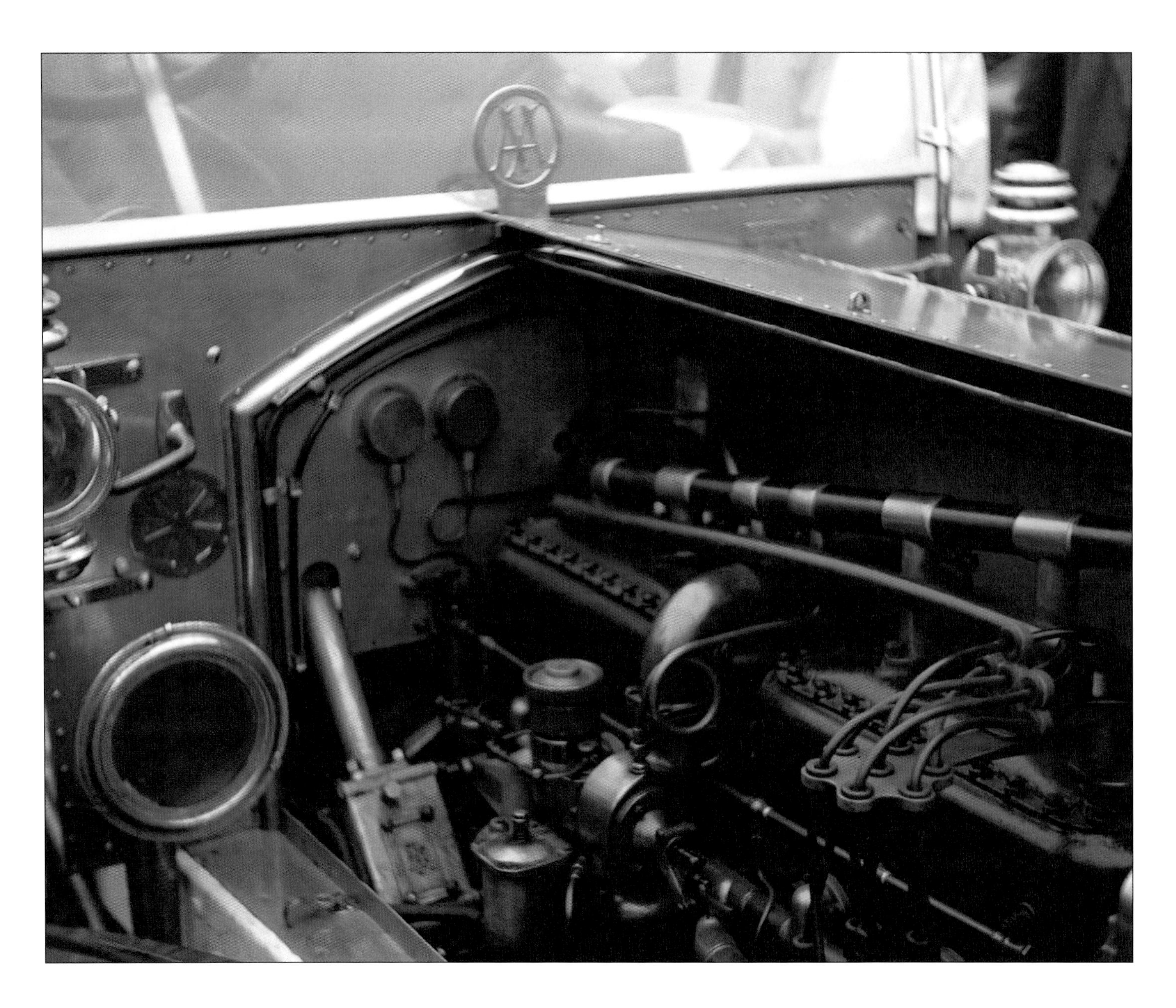

Ghost bonnet up to show the careful and thorough attention to detail of all parts and assemblies.

valves were controlled by a single camshaft, gear driven from the front of the engine. Ignition was by two systems, magneto and coil, to twin plugs and this, plus a Royce-designed carburettor, all went to give the engine its legendary flexibility.

A cone clutch took the drive back to a four-speed gearbox in which third gear was the direct drive and fourth an overdrive. While an excellent feature, this was to prove an encumbrance in an era when top gear motoring was the thing, and Rolls-Royce soon found that few customers understood that overdrive was an extra and that any comparisons with other cars should be made in third gear. In other respects the chassis followed the lines of the 1906 cars and the bodies came from specialist coachbuilders as it was not until 1945 that the firm built any for itself.

The new car went on show late in 1906 and in 1907 Claude Johnson, who became known as the hyphen in Rolls-Royce, decided to prove its worth by means of a special demonstration. For this he took the thirteenth chassis and had it fitted with a touring body finished in silver paint with silver-plated fittings. Thus was born the Silver Ghost, at first a single car but soon the model name.

Pristine 1909 Silver Ghost, just in the Edwardian era, and a car that requires a good deal of regular work to keep in this condition.

The famous Spirit of Ecstasy mascot fitted to Rolls-Royce cars from 1911. A kneeling version appeared in 1934, later changed to a shorter standing one.

ABOVE: Pictured in 1960, this Silver Ghost was then still a working car, doing a practical job.

BELOW: Silver Ghost Roi des Belges of 1909, a car requiring a great deal of polish and elbow grease to hold its shine.

For its test, the car was under RAC supervision and started off by covering 2,000 miles, was then closely examined and sent off to run from London to Glasgow, day and night, until 15,000 miles had been covered. It was then again examined, wear found to be negligible and passed with flying colours into motoring history.

Thus was established the special nature of the Rolls-Royce car, its cachet and eminence to grow over the years as it became THE car to own to demonstrate success, position and wealth. The famous slogan may have been in dispute from

This Silver Ghost is from 1911 and has a perpendicular body so typical of that time.

ABOVE: Another wonderful body from 1911, a Tulip Back with fixtures and fittings to suit.

RIGHT: A further 1911 Silver Ghost fitted with a two-seater body as a replica of a balloon car.

OPPOSITE: Interior of the 1911 Tulip Back body showing the fine panelling and other top-class workmanship involved.

ABOVE: Wire wheels in place of the artillery for this lovely 1912 Silver Ghost, the open-sided top most suited to the climate it is enjoying.

RIGHT: More fine detail work and immaculate finish on this 1912 model, showing what was expected of a body on a Rolls-Royce chassis.

time to time but no other firm spent so much time at the pinnacle of the luxury market, disdaining to build for any other. Other marques came and went, some more innovative, others more expensive, but the badge and that radiator motored on triumphantly through the years.

Although the first cars were built in Manchester, the firm moved to Derby in 1908. By then Charles Rolls was looking for new fields to conquer and turned to aviation. He was the first to fly the English Channel both ways but, tragically, the first British flyer to die in a plane crash when his Wright biplane broke up in the air in July 1910.

The Silver Ghost was the only Rolls-Royce listed from 1907 until 1922 and was built up to 1925. From 1921 it was also built in the USA at Springfield, Massachusetts, to factory specifications, only the crankshaft and a few minor

Postwar in 1919 with a Cockshoot body in a style that worked as well as any for a Rolls-Royce - and always superbly elegant.

Barker touring body on a 1922 Silver Ghost, very much the British style of the early years of that decade.

parts coming from Derby. These cars were just as well made and every bit as luxurious as their British counterparts although chassis improvements naturally lagged by a season or so, being tried at home first. Just over 1,700 were built in the USA in all.

Over the years the Silver Ghost was modified and improved in many ways, these alterations often being carried out on older cars while they were in for servicing. Rolls-Royce saw ownership as timeless and for the car to be used for a lifetime - so it became quite common for a new body to be fitted after a decade or so, if only to reflect the changing styles of the times.

Initially, from its series manufacture from 1907, the chassis was built as a short wheelbase; the long came in from 1914, and this was joined by an even longer one from 1923. The rear suspension soon altered, while the engine was enlarged late in 1909 to a 4.75in stroke and 7,428cc; in the same year a three-speed gearbox was adopted.

Drum headlights were one of the hallmarks of the American Brewster body, matched by the side lights as on this 1922 car.

In 1911 a Silver Ghost carried out an observed top-gear-only run from London to Edinburgh and a type of that name was listed alongside the standard car, it leading to the Continental model. It was also in 1911 when probably the most distinctive bonnet ornament of all - the famous 'Spirit of Ecstasy' mascot - first appeared on the top of the radiator, silver-plated in its early years, and the subject of many more tales and legends.

There were four speeds for 1913, then came the war World War 1 was the first major conflict to see the use of armoured vehicles - the Italians having used armoured cars in 1912 against the Turks in Tripolitania - and the first British ones were Rolls-Royces. Developed surprisingly by the Admiralty, equipping six Naval Air Squadrons by 1915, they were used in the early stages of the European war to defend forward airfields and rescue crews of shot-down aircraft. The Admiralty pattern Rolls-Royce was the most widely used armoured car of the war, seeing action in the Desert with Lawrence of Arabia, in Egypt, France, the Dardenelles, East Africa and in Russia where it continued to be used by British forces during the Russian civil war. Other experiments using Rolls-Royce vehicles were less successful - like the 1915 mounting of a Vickers 1-pdr pom-pom

RIGHT: Built in Springfield and fitted with a Brewster body, this 1923 model shows how well the American cars were finished.

BELOW: Derby production of the Silver Ghost ended in 1925 when this car was built, hence its front-wheel brakes.

The end of the Silver Ghost line at Springfield came in 1926, still minus front brakes, but with the lovely Brewster body as here.

with an armoured shield - but others were used successfully as staff cars or ambulances.

Civilian production restarted in 1919, four-wheel brakes did not arrive until 1924, although never in the USA, and in 1925 production of the Silver Ghost at Derby ended; the cars just as able then to run from 3 to 70mph in top gear in true Edwardian touring style. Over 6,000 had been built.

In the USA 1925 saw left-hand drive, three speeds and the purchase by Rolls-Royce of the Brewster coachbuilding firm, which was well known to the company as it had bodied many Springfield chassis. Brewster's drum-shaped headlights were a distinctive and pleasing feature that would continue to grace other Rolls-Royce cars after the production of the Silver Ghost finally ceased in 1926.

A small Rolls-Royce is a contradiction in terms, but the firm did offer the Twenty from 1922, this 1923 Cockshoot having a later radiator grill.

TWENTY & PHANTOM

1922-1929

AT the start of the roaring twenties Rolls-Royce knew that it needed a replacement for the ageing Silver Ghost and a smaller car to broaden its market appeal while remaining a luxury producer. The ultimate replacement would be the New Phantom in 1925 but before then the Twenty was introduced in 1922.

The Twenty kept to a six-cylinder engine, anything less would have been too rough for Royce, but while smaller at 3,127cc, it did feature overhead valves. It drove a three-speed gearbox with central change at first, but from 1925 had four speeds and a right-hand change.Suspension was by semi-elliptic springs, front-wheel brakes were added in 1924 and the whole car was smaller than the Silver Ghost although still bigger than many.

As with all Rolls-Royce cars of that time, the bodies came from specialist coachbuilders and were generally individual, although some firms built batches and then trimmed and finished these to the customer's requirements. The car was aimed at the owner-driver, although many were chauffeur-driven, and this brought a further variance in styles.

A limited power output restricted the performance of the Twenty, especially if the body was heavy, but a light body enabled the smooth six to cruise easily and quietly while the low stresses gave the car a very long life. It proved a popular model with over 2,900 being built up to 1929. Most had horizontal radiator shutters but these changed to vertical in 1928.

The New Phantom was announced in May 1925, later to be also known as the Phantom I after the II arrived in 1929. Various engine types were tried as alternatives to the six but in the end Royce kept to his established practice, even down to casting the cylinders in two blocks of three. Different was the one-piece cylinder head on their top and the use of overhead valves. Engine capacity went up to 7,668cc, thanks to a longer stroke, and the construction was brought up to date.

Overhead valves plus a longer stroke made for a taller engine so it was set lower in the frame behind that tall radiator which had vertical radiator shutters from the start. The chassis was much as for the Silver Ghost, including the can-

The New Phantom of 1925 took the firm on from the Silver Ghost to overhead valves and four-wheel brakes.

OPPOSITE TOP: Springfield went on to build the Phantom from 1926, this fine car dating from two years later.

OPPOSITE BOTTOM: For 1930 the Phantom II took over to continue the theme of the firm, despite the troubled financial times.

BELOW: From 1931 the Brewster factory on Long Island fitted bodies such as these to imported Phantom II chassis.

tilever rear springs, and had the highly-effective mechanical-servo for the four-wheel drum brakes. Centre-lock wire wheels were usual, although steel artillery were available, and many cars had the wheel discs common to the time.

The Phantom was a big car that maintained the Rolls-Royce cachet thanks to its undoubted performance achieved so smoothly and silently. It also kept the flexibility of the Silver Ghost that made for such pleasant motoring, the mechanical cruise control, and the high standard of workmanship for all parts and assemblies.

More than 2,200 New Phantoms were built at Derby from 1925 to 1929 and the model was also made in the USA from 1926 to 1931 with over 1,200 leaving the Springfield plant. The American version differed in more ways from the British one than had been the case with the Silver Ghost; the first change being to move the carburettor to the left side of the engine to enable it to connect easily to the left-hand-drive controls. The Americans had a centralised chassis lubrication system that relieved them of a laborious chore, but at first had to manage without front brakes although these were added to most cars later.

CG1000

Extravagant torpedo body on a 1932 20/25 model, flamboyant despite the Depression years.

Both versions changed to an aluminium cylinder head in 1928 but this was to prove troublesome although it did raise the performance. The short-chassis Continental model was faster still when fitted with a special body but most Phantoms had either the standard or long wheelbase, depending on their intended use and body type.

Changes came in 1929. By then the Twenty had become heavier and slow enough to lose its charm so the chosen simple remedy was to bore it out to 3,669cc and raise the compression ratio to change it into the 20/25 model. This livened up the performance, even if it did remain strictly touring, and the chassis gained synchromesh on the top two gears in 1932. The result proved successful in that over 3,800 were built, to be fitted with a variety of bodies whose style in general moved on with the years.

The Phantom II replaced the original late in 1929 and, although its dimensions and capacity were unaltered, the engine was much revised. A new design of

Phantom II from 1931 with Inskip body, yet another fine example of the coach builder's art.

A fine 20/25 from the 1930s, soon to become the 25/30 with more capacity and power to raise its performance.

cylinder head helped raise the power to 120bhp with more to come later with an increase in the compression ratio. The four-speed gearbox, still with right-hand change, was bolted directly to the flywheel housing and the final drive revised. The chassis had semi-elliptic rear springs in place of the cantilevers and this gave more freedom to the coachbuilders when constructing the bodies.

The Phantom remained a large, heavy car but could run to 80mph even as a limousine. The Continental chassis version built from 1932 had suspension changes and could usually run 10mph faster, thanks to the lighter body that most engines had, many among the very best of the early-1930s period.

BENTLEY ALSO
1922-1939

IN 1931 Rolls-Royce bought the Bentley company that had fallen on hard times, despite its five victories at Le Mans in the 1920s. Bentley would adopt a new image for the 1930s, but postwar moved to a much closer identity with Rolls-Royce models.

The Bentleys of the 1920s were expensive sports cars in the main, although saloons were built, and thus sold best to the sons of the wealthy who themselves rode in a Rolls-Royce. The first, a 3-litre, went on sale in 1922 and was powered by a four-cylinder, 16-valve, overhead-camshaft, long-stroke engine. In 1926 it was joined by the 6½-litre, six-cylinder model built to rival Rolls-Royce and a Speed Six derivative on a shorter chassis came in 1928. A 4½-litre had appeared in 1927 and in 1930 a small number of these were built in supercharged form, the fabled blower Bentley. Finally there came the 8-litre six in 1930 and then a 4-litre in 1931, this was last the only Bentley from the original factory not to have an overhead-camshaft engine.

A 1923 Bentley 3-litre, the first model from the firm and the start of their legend at the Le Mans races of the 1920s.

Thus, Bentley passed into Rolls-Royce hands but it was not until 1933 that a new model appeared in a newer, more refined form that gave it the name of the 'Silent Sports Car'. Built as the 3½-litre and fitted with saloon, tourer or drophead coupé bodies by the leading coachwork firms, it had a modified 20/25 engine with a new cylinder head, higher compression ratio and twin carburettors to boost the performance. The chassis had been designed for a small Rolls-Royce and was both shorter and lower so the result had a fine line and was especially suited to the sporting owner-driver of those times.

In 1930 Royce became Sir Henry but, sadly, he died in 1933 still working at the age of 70. That year the colour of the letters of the famous badge were changed from red to black, not in Royce's memory but to avoid a colour clash. It was Sir Henry's last decision for the firm.

More usual picture of a 3-litre Bentley with the hood furled; in this case a 1926 car with a Vanden Plas body.

Two years after the model's introduction, a 1928 Bentley 6½-litre with a fine touring body; the short chassis version was the Speed Six.

By the middle of the decade the effects of the Depression were wearing off and the motoring world had moved on. It was time for Rolls-Royce to update its two models and the Bentley for 1936.

Customers expected more than six cylinders in the best cars by then and Rolls-Royce responded with the Phantom III that had a new V-12 engine of 7,340cc. It was made to the expected standards, had hydraulic valve lifters, produced 165bhp and was technically ingenious, complex, expensive and troublesome if not meticulously maintained. It went into a new chassis with independent front suspension, the radiator moved forward, ahead of the axle line, and the result was a smoother car with an improved ride.

It was the same story as before, for the 20/25 added weight as the years went by, which again reduced the performance to the positively staid; once again the answer was to bore the engine out, this time to 4,257cc. There were other alterations to update the specification but in the main the model, now as the 25/30, was much as before. The car proved faster, able to cruise at 60mph, and gave owners excellent service, its maintenance easy thanks to the one-shot chassis lubrication.

ABOVE: Unusual coupé body for a 4½-litre car, not the shape associated with either the marque or the model.

BELOW: Rear quarter view of the more usual 4½-litre Bentley body and the one seen by most other drivers in the 1920s.

RIGHT: The cockpit of a 1928 4½-litre Bentley, built for performance rather than aesthetics and as brutally functional as the cars themselves.

BELOW: A rather more civilised 4½-litre Bentley thanks to its fine Vanden Plas body.

LEFT: Front of the blower Bentley with the massive supercharger driven from the crankshaft and fed by the SU carburettor under the mesh screen.

BELOW: Classic 1930 blower Bentley that was and still is the stuff of legends although the two at Le Mans in 1930 both retired.

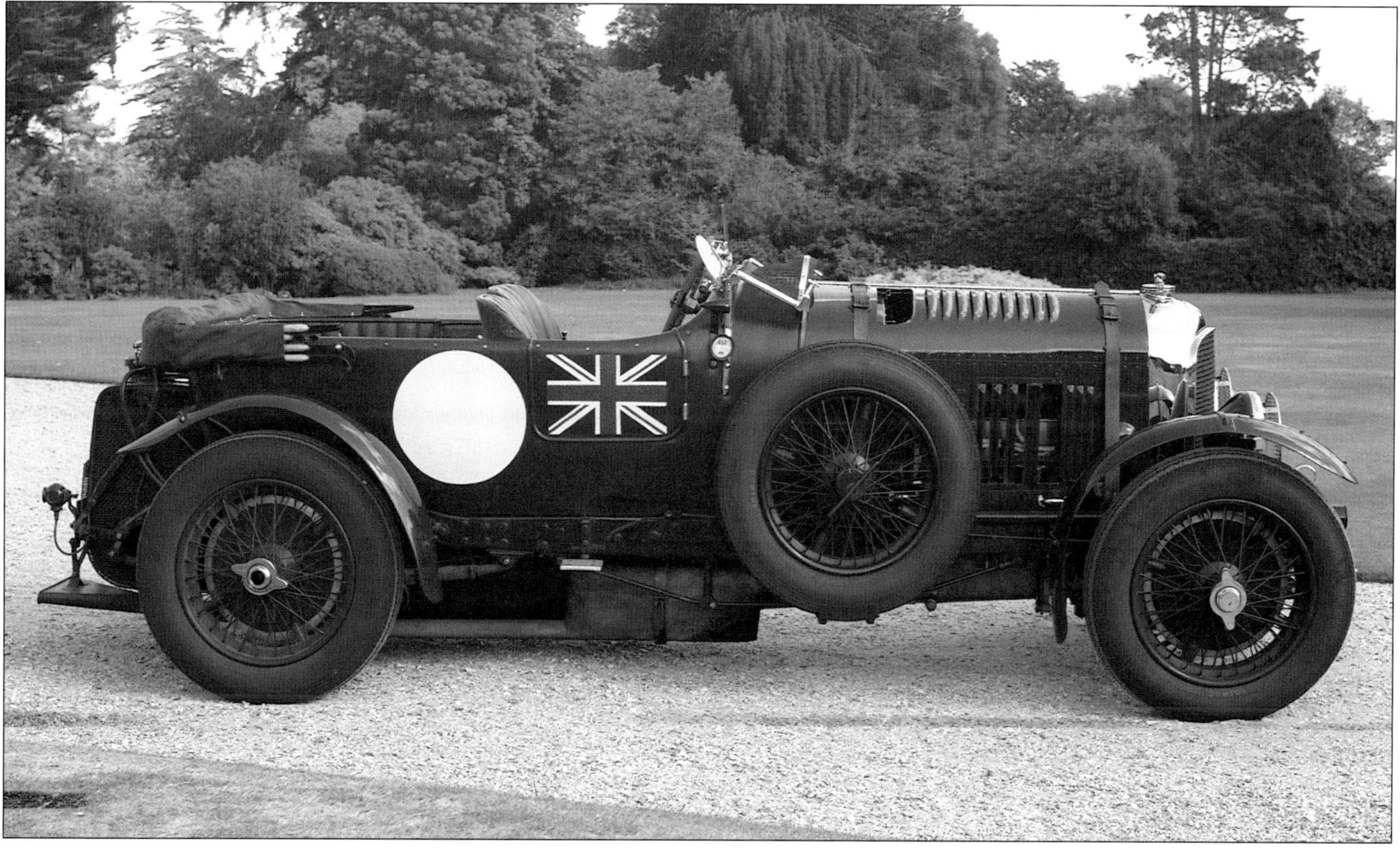

RIGHT: In later years, a 4½-litre blower Bentley out in the sun, top down and proceeding as it was always intended.

BELOW: Fine 1930 Bentley 8-litre fitted with a Mulliner saloon body to offer smooth, refined and fast cruising.

LEFT: This 1932 8-litre Bentley has a more sporting two-door body while retaining the expected performance of the marque.

BELOW: The 'Silent Sports Car' became the Bentley role under Rolls-Royce control, this a 1934 3½-litre tourer.

ABOVE: This view from above shows a different angle on the body style adopted by Bentley in the 1930s, the car a 3½-litre from 1934.

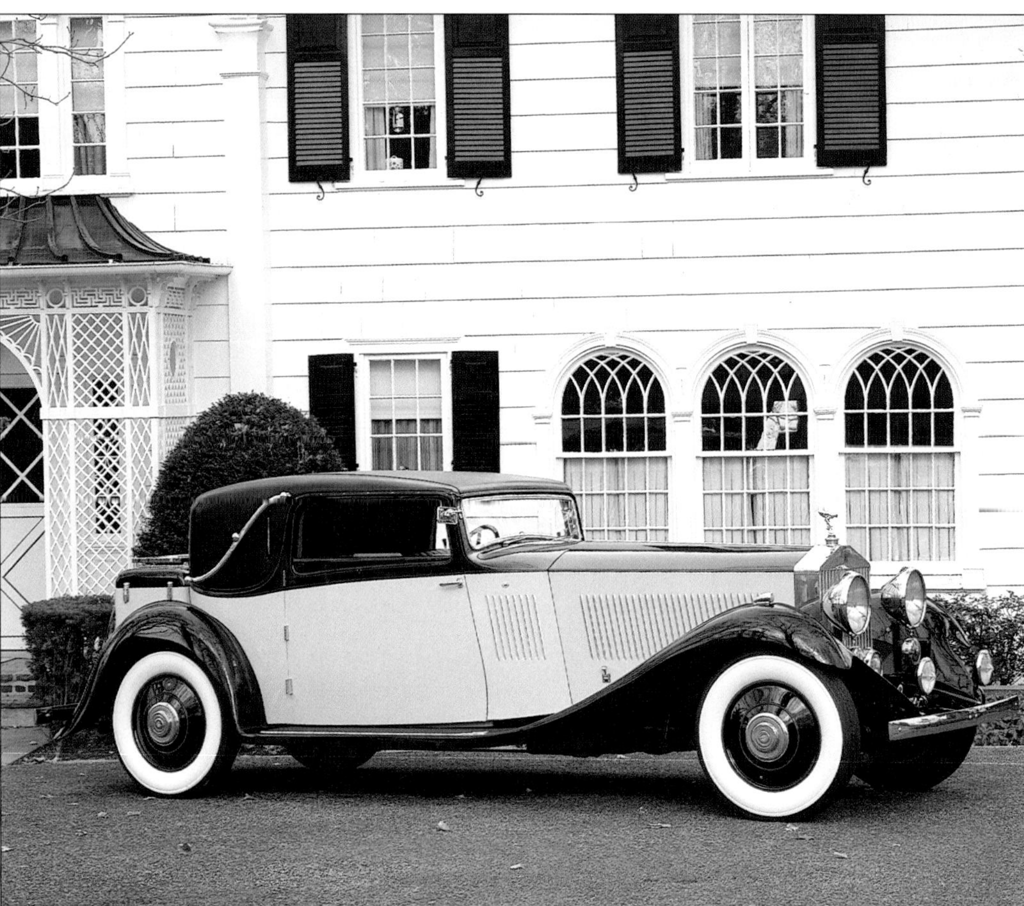

RIGHT: Up to 1935 Rolls-Royce continued to build the Phantom II - this is a fine example from 1932 with Gurney Nutting coupé body.

The Bentley change began with the 4,257cc engine being offered as an alternative but it soon became the standard and the model the 4¼-litre. The extra capacity restored the performance when under the heavier bodies, many of these came from Park Ward whose saloon was as close to a standard car as the firm went prewar. The style of all continued to be among the best of the period and gave owners fine motoring.

In August 1939 the 4¼-litre Mk V Bentley was introduced with independent front suspension and the radiator further forward. The body style was a portent of what was to come postwar but only about a dozen cars were built.

There was one more Rolls-Royce model before the war, the Wraith, that took over from the 25/30 for 1939. It kept the 4,257cc engine but gained independent front suspension, built-in hydraulic jacks, and even further refinement. The prewar production run of cars was usually fitted with a saloon or limousine body but coupés and dropheads were built.

Then came the war, the factory was involved in many other engineering matters for the duration, and afterwards car production was moved to Crewe.

Replacement Phantom III of 1938 with a fine Hooper body over its V-12 engine plus other new features.

ABOVE: The Bentley grew up to 4¼-litre in 1936, this two-door coupé from that year having something of a continental line.

RIGHT: British sports car style for 1936 by Vanden Plas on a 4¼-litre Bentley two-door convertible.

ABOVE: This style was usually known as the 'Doctor's coupé' and leaves something of the Phantom III length unused.

BELOW: Said to be an ex-Royal car, this is a 1937 Bentley 4¼-litre, typical of the style built at the time.

RIGHT: Another 1937 Bentley 4¼-litre, this time parked in Eaton Terrace, London, its turn signals a modern addition.

BELOW: The Phantom III had a complex engine that demanded minute attention to Rolls-Royce standards but offered high performance for 1938.

LEFT: Late in 1939 a handful of 4¼-litre Mk V Bentleys were built; their body style was to continue postwar.

BELOW: The last new Rolls-Royce of prewar days was the Wraith, built for 1939 to replace the 25/30 model.

WRAITH, DAWN & PHANTOM
1946-1959

AFTER the war Rolls-Royce had to change simply to survive in the luxury market. To do this the company concentrated on one engine and two chassis, the Rolls-Royce for coachbuilders and the Bentley for its own, Crewe-built, pressed-steel body. This proved an astute move for the specialist coachwork firms gradually faded away so that designs became standard, trimmed and with accessories to taste, while the two marques became as one, with only the radiators different. Sadly, this reflected in the used car values to the detriment of the Bentley whose production gradually declined until efforts in the 1980s turned this round.

The postwar models were the Silver Wraith and Bentley Mk VI. Both had a new 4,257cc engine with overhead inlet and side exhaust valves, a four-speed gearbox, hydraulic front and mechanical rear brakes, independent front suspension and semi-elliptic rear. Overall, the cars were simpler and easier to build than prewar, without some of the special and expensive features, but still setting standards for smoothness, quietness and feel.

The Silver Wraith continued to have much of the prewar style, twin Lucas headlamps beside the tall radiator surmounted by the Spirit of Ecstasy. The bodies still varied a good deal although the independent coachbuilders began to build in batches so that truly individual examples became less common. The Mk VI had far less variation although over 800 of the 5,000 produced did have special bodies including coupés and dropheads.

Just after the war and this 1946 Silver Wraith has a fine Inskip convertible body over its new engine and chassis.

LEFT: The Bentley Mk VI of 1947 shared the new engine with Rolls-Royce and here has a two-door open body based on the saloon.

BELOW: Rather more radical body style from Mallalieu on a 1948 Bentley Mk VI drophead coupé.

In 1949 the Silver Dawn was added to the range - essentially a detuned Bentley VI with a change of radiator and bonnet plus some other minor details. Thus came the standard saloon Rolls-Royce, as only a few were coachbuilt, most of these with a drophead coupé from Park Ward. The six-cylinder engine of the Silver Wraith, Silver Dawn and Mk VI was stretched to 4,566cc in 1951 when a long wheelbase Wraith became an option.

The Phantom IV appeared in 1950, marking a return to the grand car built in the smallest of numbers for royalty and heads of state. Only 18 were produced and for them the firm used a 5,675cc straight-eight engine developed for military and commercial use. The rest of the mechanics were conventional although the chassis was immense, and it carried the finest examples of the coachbuilder's art, limousines the norm.

For 1952 the Bentley Mk VI was replaced by the R-Type, a development with a longer and more sweeping rear body line. At the same time the R-Type Continental was added as a two-door fixed-head coupé, nearly all of the 200 or so produced having a body built by Mulliner in aluminium. It was aerodynamically styled and careful attention to this gave the car a top speed close to 120mph. With the advent of the R-Type, the Silver Dawn took on a similar form while remaining a detuned version of the Bentley.

Magnificent Silver Wraith from around 1951 with the stature and dignity associated with both marque and model.

ABOVE: Unusual open Royston body on a 1949 Mk VI, boasting a custom style more of the prewar SS Jaguar or Lagonda.

BELOW: The Silver Dawn was listed from 1949 and was essentially a Bentley Mk VI fitted with the Rolls-Royce radiator shell, this one from 1953.

ABOVE: Fine 1951 Bentley Mk VI with a Park Ward two-door drophead coupé body, stylish and elegant, so still the Silent Sports Car.

BELOW: The 1950 Phantom IV used by Queen Elizabeth II and fitted with a special radiator mascot of St George and the Dragon.

In 1954 the long wheelbase Silver Wraith had its engine stretched once more to 4,887cc and continued in this form to the end of the decade. The next year saw Rolls-Royce and Bentley move closer together with the advent of the Silver Cloud I, in place of the Silver Dawn, and the S1 Bentley by a change of radiator. Both used the 4,887cc engine and had automatic transmission, power steering being added from 1956.

The Continental S1 replaced the R-Type but became less distinctive as a variety of bodies were fitted. These included two-door saloons from Mulliner and Park Ward, a drophead coupé from the latter, and the Flying Spur four-door saloon. In 1957 a long wheelbase Silver Cloud I joined the lists and in this way the two ranges ran on to the end of the decade.

For 1952 Bentley replaced the Mk VI with the R-Type and introduced the Continental R two-door, fixed-head coupé.

This rear quarter view of the Continental R-Type shows its special style and form that made it such a fast car, despite its size.

ABOVE: Formal Silver Wraith from 1953, by when it was fitted with a 4.6-litre engine and mainly built in long-wheelbase form.

Below: A Rolls-Royce Silver Dawn of 1955, the standard saloon model, but owners could still have some personal touches as well as a choice of finish.

The Silver Cloud replaced the Dawn in 1955; its engine was larger, but otherwise it continued as the stock model of the decade.

A Park Ward Bentley S1 Continental convertible of 1956, the model that took over from the more distinctive R version.

ABOVE: Bentley S1 Continental convertible from 1959, its last year, finished in an interesting two-tone style.

BELOW:The Silver Wraith was built up to 1959, the year of this fine example with a body by Mulliner.

LEFT: Final year for the Bentley S1 Continental was also 1959, the year of this saloon, after which came the S2 and then the S3.

BELOW: This Silver Cloud was from 1959, the year the engine was changed from a straight six to a V-8, but there was no external indication as to which was fitted.

NEW ENGINE, NEW BODY

1959-1970

AT the end of the decade it was apparent that the six-cylinder engine could go no further thanks to the added loads placed on it by increases in car weight, power steering and air conditioning. Top speed remained fine but acceleration was being outclassed both in Europe and the USA.

The answer was a new engine, a V-8 of 6,230cc with overhead valves and well able to take all the cars on to modern times. It was not a total success at first, suffering from a number of problems that resulted in some major redesigning and a new crankcase during 1961.

In other respects the cars continued in their established form, all with automatic transmission and power steering. The Rolls-Royce became the Silver Cloud II in short and long wheelbase, the latter built as a saloon, the shorter also as a drophead coupé. The equivalent Bentleys were the S2, of which only a few had the long wheelbase, and the Continental S2 that had four-shoe front brakes and a much wider choice of bodies. These included two- and four-door saloons from Mulliner or Young, plus a drophead coupé from Park Ward. In all cases special bodies were still being built for individual customers but these became fewer as the years went by.

Old car, new engine was the theme of the Silver Cloud II that fitted the new V-8 engine, this example from 1961.

As the Silver Wraith in long-wheelbase form was dropped in 1959 a replacement was needed and this came in the form of the Phantom V. It was a new car for it used the V-8 engine and installed this in a lengthy chassis with a 12ft wheelbase. It kept to the earlier type of automatic transmission as this had the mechanical servo needed to power the drum brakes and most of the 800 built had a limousine body stretching to some 20ft. There were a few saloons and specials but this was a car for the chauffeur to drive.

In 1962 the Silver Cloud adopted quad headlights, to the dismay of the purists, and better seating to become the Series III and the Bentley followed suit as the S3. Most were built as saloons but some coupé and convertible bodies took advantage of the separate chassis that gave the coachbuilders a chance to demonstrate their skills. This was more so for the Continental S3 that continued to appear in more variety.

Late in 1965 the Silver Shadow Rolls-Royce and T-model Bentley made their debut after some nine years of design and development; lengthy, but not for a complex car that was a complete break from tradition, would have to be in production for many years, and required much testing to reach the rigorous standards of a firm whose resources were limited in comparison with the major producers.

It was perhaps inevitable that the new design would be criticised for it had to move on from the traditional values of the past and be able to adopt new techniques belonging to the future. The result was actually very successful and achieved its aims without becoming dated over a long life.

This S2 Bentley was the mate of the Silver Cloud II and, like it, was built from 1959 to 1962.

The essence of the new car was its monocoque body construction, common enough elsewhere, but spelling the end of special bodies, although by this time there were few firms left able to do this anyway. Coupés and convertibles were built, but by altering and bracing the stock saloon body panels to suit. As a result, there was less variation of style but the four headlamps remained a feature.

Under the body went the V-8 engine with minor modifications, this driving automatic transmission as usual. In other respects there were many changes with independent suspension front and rear, a self-levelling hydraulic system and disc brakes all round. This added up to much new design work and a complex car, especially with the various hydraulic systems of both brakes and ride height.

The Silver Shadow was built in two wheelbases, the Bentley in one and in much smaller numbers as the market found that it lost value far more quickly, despite being a Rolls in all but radiator shell. It was much the same with the coachbuilt saloons and convertibles, most options being available but the numbers were small.

The replacement for the Silver Wraith limousine was the Phantom V built from 1959; this is a 1963 car, large and V-8 powered.

LEFT: A further Silver Cloud, this time a III from 1963, hence the quad headlights that were not to the taste of all owners.

BELOW: Bentley S3 of 1963 in its long-wheelbase form and, as with the Silver Cloud III, fitted with quad headlights to complete an altogether imposing car.

Discreet two-tone elegant finish on a 1964 Silver Cloud III in true Rolls-Royce style.

Performance and handling of the new cars was excellent and they established a format that would continue for many a year. Good maintenance was an essential but resulted in a very reliable machine that continued to be built to the highest standards. Alterations appeared in all areas, often as technology moved on for items such as electric generation, radial tyres or central door locking.

In 1968 the Phantom VI replaced the earlier version with little real alteration and continued as an exclusive limousine. It kept to the original V-8 engine for some years but the Silver Cloud and T-model had their engine capacity increased to 6,750cc during 1970.

LEFT: Continental S3 Bentley of 1963 that continued the two-door, fixed-head coupé style as before.

BELOW: This 1968 Silver Shadow had the standard four-door saloon body, a most successful design that did not date over its long life.

ABOVE: A Silver Shadow from 1968 fitted with a two-door drophead coupé body by Mulliner Park Ward, its top power-operated.

BELOW: Bentley T1 from 1966, the model that along with the Silver Shadow, took the firm to monocoque construction and no separate chassis.

ABOVE: The magnificent and lengthy Phantom VI that took over the exclusive limousine role for a decade from 1968.

LEFT: Interior of a 1970 Phantom VI that gives some idea of the style, fittings and standard of finish lavished on these cars.

CORNICHE & CAMARGUE, SPIRIT & SPUR

1971-1996

A new name appeared in 1971 for the coupé and convertible versions of the Silver Shadow - the Corniche. For a few years it was also built as a saloon, until the Silver Shadow was replaced, but, remarkably, it was to run on for 23 years with little external alteration, a real tribute to its fine lines that never became dated. Up to 1984 it was also built as a Bentley but in very small numbers before it was renamed the Continental to run on with the Corniche to the end.

The new decade was a time of much trauma for the firm as the aero-engine division went bankrupt and for a time the future of the cars was in some doubt. However, Rolls-Royce Motors was formed and was able to weather the storm although only one new model appeared in the mid-1970s.

This was the Camargue that had a Pininfarina designed two-door body built by Mulliner Park Ward on the Silver Shadow floorpan and mechanics. While not to the taste of all, it was individual and featured a number of fittings that would appear in time on other models. It was heavier than a Shadow but the shape allowed it to run faster and while production was little more than 500 over 11 years, it was a successful model.

Silver Shadow as for 1973, continuing as a car of dignity and style with speed, silence and great comfort.

Changes came for 1978 with a facelift for the Silver Shadow, that became the Mark II and the Bentley T2. The long wheelbase Shadow became the Silver Wraith II while the Phantom VI was finally fitted with the larger 6,750cc V-8 engine.

More changes came in 1980 when the firm became part of the Vickers group as it is to this day. Vickers was an excellent firm to absorb the company because of its historic involvement with and use of Rolls-Royce engines for its armoured fighting vehicles - especially the Centurion main battle tank family for so long the workhorse of the British army. It was the start of a decade when the Bentley name would be promoted more, sales of the marque having sunk to one for every 20 Rolls-Royce. The Corniche, Camargue and Phantom continued but the others were given a new style and names.

The result was the Silver Spirit, the long wheelbase Silver Spur and the Bentley Mulsanne. All kept the V-8 engine but the body was wider and lower, lost some of the angular style of the Shadow and looked bulkier, perhaps reflecting its increase in weight. However the build quality and refinement remained as high as ever, while improvements evolved year on year using new materials and techniques as they came along.

This 1970 Silver Shadow had a two-door saloon body from Mulliner Park Ward, the style improved by the nip in the side between the door and the rear wheel.

The Corniche convertible was introduced in 1971 and ran right into the 1990s with minimal style changes, this car from 1975.

In 1982 the Mulsanne Turbo was introduced and its turbocharged engine emphasised the performance image of the marque, and perform it did, its 135mph top speed limited by the tyres available. A long wheelbase Mulsanne appeared at the same time but the recession hit the firm hard, sales halving in two years. The Bentley response was to rename their Corniche the Continental, and the Mulsanne Turbo the Turbo R while introducing a slightly cheaper model. This was the Eight that featured a wire mesh radiator grille to remind people of the glories of Le Mans, a number of minor changes, and a price tag £10,000 down from the Mulsanne. Built up to 1992, it sold well.

The Rolls-Royce models continued year on year, the V-8 engine changing to fuel injection around 1987 when the Corniche became the Mark II and the Mulsanne the Mulsanne S. For 1989 it was Corniche III, Silver Spirit II and Silver Spur II while 1991 brought the Corniche IV.

By 1992 the Phantom VI had run its course and its replacement was the Silver Spur Touring Limousine that continued the theme in a magnificent manner. The existing long wheelbase car was extended a further 2ft, the roof raised and Mulliner Park Ward built a new body equipped and finished in the most lavish way to the highest standards. The result was just as exceptional as always.

ABOVE: Bentley listed its Corniche model from 1971, when this car was made, to 1984, after which it was renamed the Continental

LEFT: New in 1975 was the Camargue that had a Pininfarina body design executed by Mulliner Park Ward on a Silver Shadow base.

LEFT: Soon due for a facelift, this is a 1976 Silver Shadow I, the long wheelbase version was later to adopt the Silver Wraith name.

ABOVE: From late 1977 the Silver Shadow II replaced the earlier car, having had a number of specification and styling changes

RIGHT: To match the Shadow, the Bentley T2 had changes for the late 1970s but sold in much smaller numbers.

RIGHT: A view from above of a 1978 Corniche convertible with the top down and buttoned in place.

ABOVE: The Silver Spirit replaced the Shadow from 1980, its body having a somewhat heavier and bulkier line.

BELOW: A further view of the Silver Spirit of the 1980s, bulky, but slim from behind the wheel and no problem in heavy traffic.

There was also a new Bentley for 1992, the Continental R that combined the turbocharged engine with a fine two-door coupé body built for Bentley alone. It recalled the distinctive R-Type of 1952, was expensive, but a high performer. Further new Bentley models appeared in 1993, the entry-level Brooklands in standard and long wheelbase forms replacing the Mulsanne S.

The Silver Spirit III and Silver Spur III replaced the older models for 1994, further refined thanks to a massive investment programme. There was also the Flying Spur, a more powerful version, and from this came the limited edition Bentley Turbo S that built on the success of the Turbo R with the top speed up to 155mph. A concept Bentley, the Java, was shown at a motor show, this using a 3.5-litre V-8 engine and having many advanced features for evaluation.

During 1994 the Corniche came to the end of its long production run. There had already been an Anniversary Edition in 1992 to celebrate 21 years but the firm decided that it really was time to close the order books.

There was a much altered range for 1996 when body styles were revised, interiors improved, engine outputs raised while fuel consumption was reduced. There was no alteration in the quality and refinement of the fixtures and fittings, while the finish was further protected by the lengthy and advanced treatment processes used.

Models simply had names without the numerical suffix so the three Rolls-Royce cars were the Silver Spirit, longer Silver Spur and much longer Limousine, all powered by the well-developed V-8 engine. There were four Bentleys, the Brooklands saloon using the same engine, while the other three had the turbocharged version. Of these, the Turbo R and Continental R kept to the existing

The Silver Spur had a longer wheelbase than the Spirit and was thus an impressive car, this example from 1987.

ABOVE: Bentley Mulsanne from 1981, the companion to the Silver Spirit and Silver Spur models.

LEFT: A stretched Silver Spur limousine on Oregon plates, the result still keeping its style and looks.

format but the Azure took the place of the Corniche. It differed in that the two-door body had a fully-powered hood so could go from open tourer to closed, four-seater coupé at the touch of a button. The roof stowed under the rear deck when not in use.

That year the firm reminded itself of Henry Royce's words, 'Take the best that exists and make it better' that reflected the outlook that had kept Rolls-Royce at the top of the luxury car industry for so long, the only company whose sole product was such cars.

ABOVE: Typical Rolls-Royce fascia finished in burr walnut and fitted with switches and controls in the fashion of the company.

BELOW: Mulsanne Turbo Bentley that was introduced for 1982 to offer more performance from the established model.

To avoid confusion, the Bentley Corniche model was renamed the Continental for 1984 while keeping the style of the type.

For 1985 the Turbo R replaced the Mulsanne Turbo, essentially a change of name only, and this example is from three years later.

Under the bonnet of a Bentley Turbo R to show how the engine and all the ancillaries were packed in.

RIGHT: The Bentley Eight had a reduced price and a radiator grill that hinted of past glories at Le Mans.

OPPOSITE: Safety regulations dictated that the Spirit of Ecstasy should not be a danger so it was arranged that it dropped out of sight when any threatened.

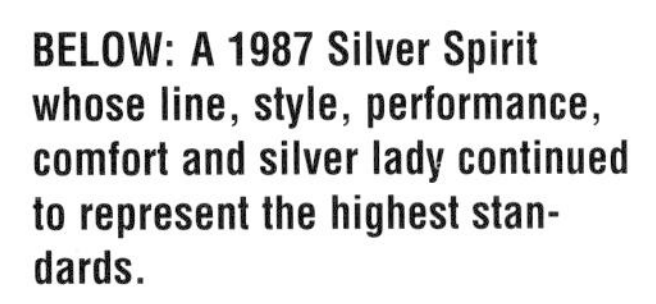

BELOW: A 1987 Silver Spirit whose line, style, performance, comfort and silver lady continued to represent the highest standards.

ROLLS
ROYCE
200 RRR
ARNETT of BOURNEMOUTH

ABOVE: Corniche II of 1987 continued the well established theme of the luxury, four-seat convertible that proved so popular.

BELOW: The Mulsanne S replaced the earlier Bentley model in 1987, the year that fuel injection went on the V-8 engine.

LEFT: In 1989 the Corniche III took over and its centre console and interior were as shown here.

BELOW: The Silver Spirit II of 1989 represented a further development of the series with discreet improvements.

ABOVE: Rolls-Royce continued to offer the longer wheelbase saloon as the Silver Spur II of 1989.

RIGHT: In 1991 the Corniche IV appeared, there being few external alterations and no real need for them.

The Silver Spur Touring Limousine took over the duties of the old Phantom VI in 1992, longer and higher than the stock Spur.

New for 1992, the Continental R Bentley revived memories of a 1950s car and was not matched by a Rolls-Royce model.

The Bentley Brooklands of 1993 replaced the Mulsanne S and was listed in two wheelbase lengths.

In 1994 the Silver Spirit III replaced the series II, further improved thanks to massive investment.

ABOVE: The Silver Spur III matched the Spirit III while retaining its extra length and own style and sales niche.

RIGHT: The Flying Spur was an additional model for 1994, fitting a more powerful engine for greater performance.

RIGHT: From the Flying Spur came this Bentley Turbo S of 1994, very fast and very select.

In 1994 the Corniche was finally dropped after a 23-year run, this car one of the series IV built from 1991.

Revised for 1996, this is the Silver Spirit that remained as close to a standard Rolls-Royce as was feasible

The longer Silver Spur continued as the stretched Spirit for 1996, just as before and just as elegant.

RIGHT: The 1996 Bentley Brooklands remained based on the Silver Spirit and used the same V-8 engine and mechanics.

RIGHT: Turbo-charging was used for the 1996 Bentley Turbo R model to increase its performance.

BELOW: Bentley Continental R for 1996 that kept to the two-door, fixed-head coupé body style from the past and fitted the turbo engine.

ABOVE: New for 1996, the Bentley Azure open tourer that had a fully powered top and used the turbo engine.

BELOW: The interior of the 1996 Silver Spur, much changed from the 1907 Silver Ghost but still reflecting the aims of the founders.

ROLLS-ROYCE AND BENTLEY MODELS

1904-05 10hp, 2 cylinder
1905 15hp, 3 cylinder
1905-06 20hp, 4 cylinder
1905-07 30hp, 6 cylinder
1907-25 40/50 Silver Ghost
1921-26 US 40/50 Silver Ghost
1922-29 Twenty
1925-29 Phantom I
1926-31 US Phantom I
1929-35 Phantom II
1929-36 20/25
1932-35 Phantom II Continental
1936-39 Phantom III
1936-38 25/30
1938-39 Wraith
1946-51 Silver Wraith
1949-51 Silver Dawn
1950-56 Phantom IV
1951-52 Silver Dawn, 4.6 litre
1951-53 Silver Wraith, 4.6 litre
1951-54 Silver Wraith, LWB
1952-55 Silver Dawn, 4.6 litre
1954-59 Silver Wraith, LWB
1955-59 Silver Cloud I
1957-59 Silver Cloud I LWB
1959-62 Silver Cloud II
1959-68 Phantom V
1962-65 Silver Cloud III
1965-70 Silver Shadow I
1965-71 Silver Shadow I Coupe
1967-71 Silver Shadow I Convertible
1968-78 Phantom VI
1969-70 Silver Shadow I LWB
1970-77 Silver Shadow I, 6.7 litre
1970-77 Silver Shadow I LWB, 6.7 litre
1971-86 Corniche I
1975-86 Camargue
1977-80 Silver Wraith II
1977-80 Silver Shadow II
1978-92 Phantom VI, 6.7 litre
1980-89 Silver Spirit
1980-89 Silver Spur
1987-88 Corniche II
1989-90 Corniche III
1989-93 Silver Spirit II
1989-93 Silver Spur II
1991-94 Corniche IV
1992 Corniche Anniversary
1992-95 Silver Spur Touring Limousine
1994-95 Silver Spirit III
1994-95 Silver Spur III
1994-95 Flying Spur
1996 Silver Spirit
1996 Silver Spur
1996 Limousine

BENTLEY

1922-29 3 litre
1926-30 6½ litre
1927-31 4½ litre
1928-30 Speed Six
1930-31 blown 4½
1930-31 8 litre
1931 4 litre
1933-36 3½ litre
1936-39 4¼ litre
1939 Mk V
1946-51 Mk VI
1951-52 Mk VI, 4.6 litre
1952-55 R-Type
1952-55 Continental R-Type
1955-59 S1
1955-59 Continental S1
1959-62 S2
1959-62 Continental S2
1962-65 S3
1962-65 Continental S3
1965-70 T1
1970-77 T1, 6.7 litre
1971-84 Corniche
1977-80 T2
1980-87 Mulsanne
1982-84 Turbo
1982-87 Mulsanne LWB
1984-92 Eight
1984-94 Continental
1985-93 Turbo R
1985-93 Turbo R LWB
1987-92 Mulsanne S
1992-95 Continental R
1993-95 Brooklands
1993-95 Brooklands LWB
1994-95 Turbo S
1996 Brooklands
1996 Turbo R
1996 Continental R
1996 Azure